a visit to the NATURAL HISTORY MUSEUM

Created by The Child's World

Photographed on location at Field Museum of Natural History, Chicago, USA

For days Mrs. Kranz and her class have been talking about their visit to the Natural History Museum. Now that they are here, the excited children want to hurry inside. Mrs. Kranz checks to be sure everyone has arrived. Then she says, "Let's go in."

Photographed on location at Field Museum of Natural History, Chicago, USA

In the Great Hall, Mrs. Kranz stops near the Kodiak Bear.

"Look at the big bear," says Jazzy.

"He really is big," says Laura.

Courtesy of Field Museum of Natural History, Chicago, USA, transparency #GN84727c

"I think everything is big," says Jessica, looking down the Great Hall.

"May we go see the elephants up close?" Luke asks. "I like elephants a lot."

Mrs. Kranz says yes.

Courtesy of Field Museum of Natural History, Chicago, USA, transparency #GN83723

"Nowadays people can see live animals at zoos," Mrs. Kranz says. "But elephants are having a hard time surviving in the wild. If all the elephants ever disappear, as the dinosaurs did, people can still see them here. That's one of the reasons there are natural history museums."

"I'm glad there are," says Luke.

Courtesy of Field Museum of Natural History, Chicago, USA, transparency #GEO-84515c

Photographed on location at Field Museum of Natural History, Chicago, USA

The children get to see a dinosaur next. He's in the Great Hall too. His name is Albertosaurus.

Courtesy of Field Museum of Natural History, Chicago, USA, transparency #A-108441c

As the children walk along, they pass some tall totem poles. They like looking at them. So Mrs. Kranz says they will visit Indian exhibits next.

Courtesy of Field Museum of Natural History, Chicago, USA, transparency #A-108729c

Soon Mrs. Kranz and the children find a whole forest of totem poles. "These were made by Indians in Washington State, Canada, and Alaska," says Mrs. Kranz.

Photographed on location at Field Museum of Natural History, Chicago, USA

Some totem poles, especially big house posts, are colorful. Phillip's dad, Mr. Noll, asks, "Can you find the ravens and the sea grizzly bears, holding humans? They are supposed to be on these poles."

A sign helps the children find them.

Courtesy of Field Museum of Natural History, Chicago, USA, transparency #A-109209c

On their way to the Pawnee Earth Lodge, the class passes many costume displays.

"This is fun," says Laura. "Whenever I walk up to a glass case, the lights turn on. If I back up, they go out again." The children try to fool the lights, but they can't.

Courtesy of Field Museum of Natural History, Chicago, USA, transparency A#106901

When they get to the Pawnee Earth Lodge, the gate is closed. The children can look over, but they have to wait for a program to start before going in. They only wait a few minutes.

Courtesy of Field Museum of Natural History, Chicago, USA, transparency #WB-2, photo by William Burlingham

Once inside, the children hear songs and stories that Pawnee Indian children once enjoyed. And they touch rattles and saddles, arrows and tools.

Photographed on location at Field Museum of Natural History, Chicago, USA

Leaving the Indian exhibits, the children start toward the special "Sizes" exhibit. It will help them learn that people, animals, and other things come in different sizes.

Photographed on location at Field Museum of Natural History, Chicago, USA

In "Sizes," Steven and Ethan help Jon try on some shoulder pads that were worn by William Perry of the Chicago Bears.

Photographed on location at Field Museum of Natural History, Chicago, USA

Chantelle and Jazzy try on the largest-sized jeans Levi Strauss makes. "They don't fit you very well," says Missy, grinning.

"Those small ones won't fit you either," says Chantelle. The children giggle.

Photographed on location at Field Museum of Natural History, Chicago, USA

Next Chantelle, Jazzy, and Melissa discover a bigger-than-life table. Chantelle's mother sits down. Her feet dangle. "This is just as it was when I was a child," she says. "My feet don't touch the floor."

"Mine didn't either," Chantelle tells her.

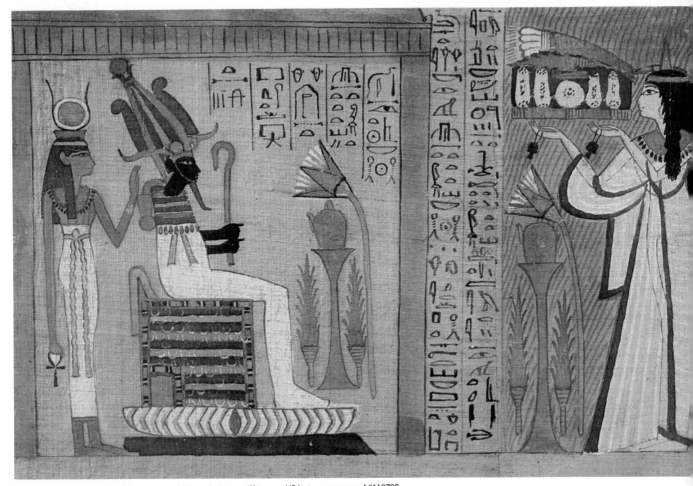

Courtesy of Field Museum of Natural History, Chicago, USA, transparency A#110708

The group visits the Egyptian exhibits next. First they see part of a scroll.

"This scroll is made of papyrus," Mrs. Kranz says. The children can't read the strange writing, but they enjoy looking at the pictures.

"Let's find the mummies next," says Mrs. Kranz.

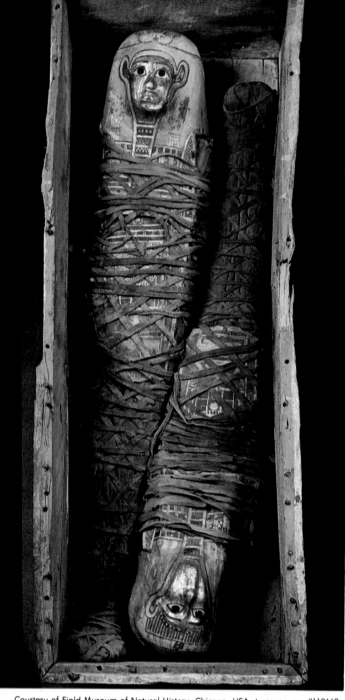
Courtesy of Field Museum of Natural History, Chicago, USA, transparency #110660c

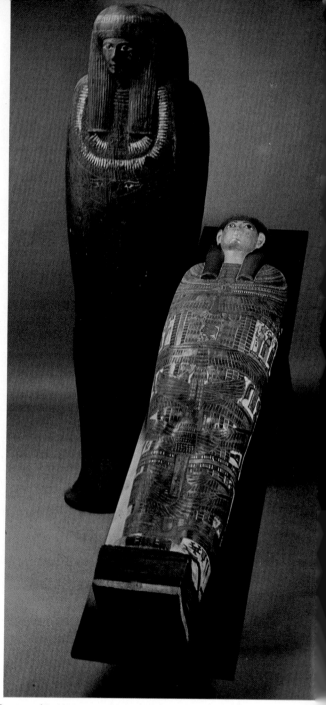
Courtesy of Field Museum of Natural History, Chicago, USA, transparency #1107

Her class discovers twin mummies in a wooden box, mummies in colorful cases, and even . . .

Courtesy of Field Museum of Natural History, Chicago, USA, transparency #A269T

Courtesy of Field Museum of Natural History, Chicago, USA, transparency #A-108670c

the most fascinating mummy of all—Mummy Harwa.

The mummies are near the False Door from an Egyptian burial tomb.

Mrs. Kranz says that a false door looks like a door, but really isn't. It's there for the soul to use, because the Egyptians believed that the soul could come and go from the grave.

Photographed on location at Field Museum of Natural History, Chicago, USA

Courtesy of Field Museum of Natural History, Chicago, USA, transparency #Z-2T

Leaving "Egypt," the children discover a big, stuffed gorilla. His name is Bushman.

"The sign says he weighed 550 pounds. He was 6 feet, 2 inches tall," says Steven.

"Yes," says Steven's dad. "Bushman was a favorite at the zoo. People were sad when he died. But they were glad he could come here."

Photographed on location at Field Museum of Natural History, Chicago, USA

Nearby, they stop to look at some seashells and coral. "The one looks just like a brain," says Johanna.

"That's why it is called brain coral," Jon answers. Jon listens for the sound of the sea. Johanna watches, waiting for her turn.

Courtesy of Field Museum of Natural History, Chicago, USA, transparency #B-83007c

Photographed on location at Field Museum of Natural History, Chicago, USA

"These plants look almost real," says Ethan's mother, as they stop to look at plants from other lands. The children like that about the museum. Lots of exhibits look real.

Courtesy of Field Museum of Natural History, Chicago, USA, transparency #G-37T

At last they come to Dinosaur Hall. "Look, more bones," says Micah. And there are.

The class agrees that their favorite animal is the 72-foot long Apatosaurus, but...

Courtesy of Field Museum of Natural History, Chicago, USA, transparency #GEO-84564c

the Irish deer . . .

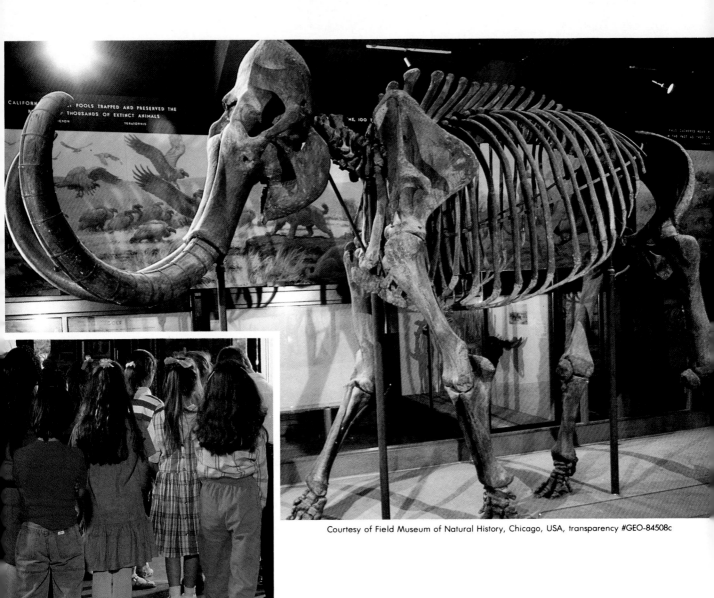

Courtesy of Field Museum of Natural History, Chicago, USA, transparency #GEO-84508c

Photographed on location at Field Museum of Natural History, Chicago, USA

and the mammoth are interesting to see too. They all lived very long ago.

Photographed on location at Field Museum of Natural History, Chicago, USA

At last the children come to a special part of the museum—The Place for Wonder. "It's okay to touch things here," says Luke. And they do.

Photographed on location at Field Museum of Natural History, Chicago, USA

Everyone pets the polar bear. But Jamar says he's glad it is stuffed. Tina agrees.

Photographed on location at Field Museum of Natural History, Chicago, USA

Jeremy and Katie try on some costumes. Katie, with her musical instrument and her "Happy Coat," is trying to talk Jeremy into doing a dance.

"Too late," says Mrs. Kranz, "because now it is time for us to be going."

Photographed on location at Field Museum of Natural History, Chicago, USA

"I hope I get to come again someday," says Steven.

"So do I," Mrs. Kranz answers. "This museum is a wonderful place to visit."